12/95

PENNY RAIFE DURANT lives in Albuquerque, New Mexico, with her husband Omar and her son Adam. A former elementary school teacher and preschool teacher and director, she is the author of several science books for children and one young adult novel, *When Heroes Die.*

BUBBLEMANIA!

PENNY RAIFE DURANT

Illustrations by Gary Mohrman

AN AVON CAMELOT BOOK

BUBBLEMANIA! is an original publication of Avon Books. This work has never before appeared in book form.

AVON BOOKS
A division of
The Hearst Corporation
1350 Avenue of the Americas
New York, New York 10019

Text copyright © 1994 by Penny Raife Durant
Interior illustrations by Gary Mohrman
Illustrations copyright © 1994 by Avon Books
Published by arrangement with the author
Library of Congress Catalog Card Number: 93-38274
ISBN: 0-380-77373-2
RL: 5.0

Library of Congress Cataloging in Publication Data:
Durant, Penny Raife.
    Bubblemania! / Penny Raife Durant.
        p.   cm.—(Avon Camelot book)
    1. Bubbles—Juvenile literature.   2. Surface tension—Juvenile literature.   3. Bubbles—Experiments—Juvenile literature.   4. Soap bubbles—Experiments—Juvenile literature.   [1. Surface tension—Experiments.   2. Bubbles—Experiments.   3. Experiments.]
I. Title.
QC183.D87   1994   93-38274
530.4'275—dc20   CIP   AC

First Avon Camelot Printing: May 1994

CAMELOT TRADEMARK REG. U.S. PAT. OFF. AND IN OTHER COUNTRIES, MARCA REGIS-TRADA, HECHO EN U.S.A.

Printed in the U.S.A.

OPM   10   9   8   7   6   5   4   3   2   1

For my mother, Patty Raife,
with lots of love

ACKNOWLEDGMENTS

My thanks go to the third period gifted students in Mike Osborn and Jeanetta Carmignani's Humanities Seminar for their creative help, and to my critique group for so very much.

TABLE OF CONTENTS

INTRODUCTION

Bubbles are fun for everyone! Little kids love blowing bubbles. So do big kids. So do parents and grandparents.

This book is all about bubbles. Large bubbles. Small bubbles. Lots of bubbles. Bubbles in your soda pop. Bubbles in your bathtub. Bubbles to blow. Bubbles to sculpt. Bubble makers. Bubbles and art. Bubbles and toys. Bubbles and your neighborhood.

Some scientists study bubbles. They have expensive equipment in their laboratories. You won't need much equipment to begin learning about bubbles. All you'll need can probably be found in your kitchen, in your garage, or at the grocery or drugstore.

Be sure you ask before you borrow things. Or, better yet, save up some allowance and buy your own ingredients. Either way, once you get started, you will want to spend lots of time with bubbles. Your friends might feel left out if you're not spending time with them, so invite them to experiment, too! You can get the whole neighborhood or maybe your school involved in a bubble day.

First, we'll learn about surface tension. Then we'll experiment with soap films. Eventually, we'll learn how to blow gigantic bubbles. Don't worry! There will be fun experiments and activities to try all along the way.

1

BUBBLE TROUBLE: SAFETY

Before we begin, we need to talk about safety. While none of the ingredients in the bubble formulas will hurt you in small amounts, it is always a good idea to be careful. Try not to get bubble liquid in your eyes. If a bubble does pop in your eye, rinse your eye with clear water.

Be especially careful with bubbles and little children. When you add things to the bubble formulas that make them smell as if they'd taste good, you need to make sure little kids don't try to drink the solution. If a bubble pops on your face or if your dog bites one, it won't hurt. It won't taste good, either. It will taste soapy, but it should rinse right away.

Another thing to remember is that soapy water can make things slippery! Unless the weather is bad, it's best to do bubble blowing outside, on a surface that doesn't get slippery, like your driveway or the sidewalk. Even grass can get slick from bubble liquid that spills and from popped bubbles. Just be careful.

Last of all, bubbles can be very messy. That's one of the things that makes them fun. If you are inside and experimenting with bubble frames, domes, and sculpting bubbles, be sure to cover your table with newspapers or an old plastic tablecloth. Cover the floor around the table with newspapers, too. Don't do bubbles where there is carpet, and be very careful of bubble liquid that has spilled on a tile or vinyl floor. It will be very slippery!

WATER'S INVISIBLE WALLS

Have you ever seen an insect stand on the surface of a lake or pond? Did you wonder why it didn't fall into the water and drown? Why does a leaf float on the top of water while a rock falls beneath the surface?

To understand why bubbles form, we have to understand that water has *surface tension*. This means that the top of water behaves as if it had a skin, or covering. It is this skin that lets an insect or leaf stay on top. This skin isn't terribly strong because a rock may pass right through it.

Watch some water drip from a faucet or make some drops yourself with a plastic straw. Adjust the amount of water coming through the straw or faucet to form drips very slowly. The water collects, hanging on the edge until enough has joined together to make a drip. We think of a drip as being rounded with a pointed top. Watch your drips. Can you tell what shape they are just before they hit the ground? They are perfect spheres. The surface tension pulls the pointed end into the rest of the drop. Can you see the spheres? No? Don't feel bad. They are hard to see because they drop so quickly. Photographers have captured drips in photos, however. That's one way we know they're spheres.

If there are no other forces acting on water, it forms a sphere. However, on earth water has to deal with gravity, and gravity exerts a tremendous force on water. That's why ponds, lakes, and rivers all appear to be flat. But, are they really flat?

 Take a small piece of waxed paper. Using an eyedropper or a straw, drop several drops of water on the waxed paper. Look at them with a magnifying glass. What shape are they? Take a toothpick and try to flatten them. What happens? Blow them with a straw. Can you blow them all together? Can you blow a big drop into smaller drops? Look at them again with the magnifying glass. What shape are they now?

 Take a clear glass. Fill it full of plain water. Keep adding until you know it is full. Take an eyedropper. Add more water, one drop at a time. Look at the glass from the side. Keep adding more drops until the top of the water is above the top of the glass.

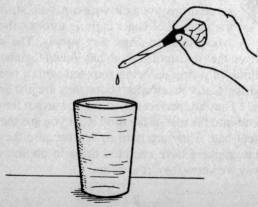

Why doesn't the water spill? The surface tension holds the water at the top of the glass together, allowing it to fill beyond the rim of the glass.

Slide a penny down the edge of the glass. Does the water overflow? How many pennies do you have to add to make it overflow? Have a contest with your friends. See who can add the most pennies without spilling any water.

Take a clean, dry glass. Add enough water to make it three-fourths full. Set a small cork on the surface. Does it stay in the middle of the glass? Look at the surface from the side. Do you see the rounded top of the water? What effect does this have on the cork?

Remove the cork. Take your finger and dip it into the water. Run your wet finger around the edge of the inside of the glass.

Add the cork again. What happens? Look at the glass from the side. Do you see the rounded top of the water? What effect does your wetting the sides of the glass have on the surface of the water? Where does the cork stay?

Try letting the glass sit overnight without moving it. Look at it again the next day. What happens? Does the surface tension pull all the water back together?

CAN YOU FEEL THE SURFACE TENSION OF WATER?

 Fill your kitchen sink half-full of water. Take a spatula. Strike the water with the flat side of the spatula. What happens? Try it again with the thin side. What was the difference?

Try it again. This time, use your hand. Put your hand into the water with your palm flat against the top of the water. Do you feel the resistance? Now try again with the side of your hand entering the water first. What was the difference?

Have you ever watched a diving competition? Really good divers barely splash the water when they go under. They try to enter the water straight, with their bodies following the hole in the surface tension their hands made. Have you ever jumped into a swimming pool with your body flat? That is called a "belly flop." If you've done it once, you'll remember for a long time. It hurts! The surface tension of the water in the pool gives way to your body. But not until it has resisted.

WATER "BEADS UP" ON
SOME SURFACES

Look at your car after a rain. Does the water stand in little pools? If so, you have a good wax job on the car. Try spraying water on different surfaces. Check with an adult before you spray anything inside. Spray the countertop in the kitchen, the bathroom sink, vinyl flooring, cement. Use a rag to dry off the surfaces after you have tested them.

The water that is repelled by the surfaces "beads up," or forms drops. If the surface absorbs water, the water sinks into the surface, and no drops form.

Take a smooth plastic or glass plate—you'll need a surface that does not absorb water. Add a few drops of food coloring to one-fourth cup of water. Pour half the water onto the plate. Take some rubbing alcohol. Pour about one-fourth cup on top of the colored water. What happens? The alcohol and water do not mix. The water, when it comes into contact with the alcohol, pulls back. The surface tension tightens. Do it again after you rinse and dry the plate. This time, try to pour the alcohol into the center of the colored water. Try it again with just one tablespoonful of each liquid. If you leave it for a few minutes, will they eventually join?

IS THERE TENSION BELOW
THE SURFACE?

 Take an artist's paintbrush, the kind you get in a set of watercolors. Look at the dry bristles of the brush. Dip the brush in water. Observe the bristles again. Do they cling together? Put the brush back into a clear glass of water. Are the bristles still clinging?

Put the brush back into the water. Slowly raise it. Look at the brush from the side of the glass when the bristles are halfway out of the water. Can you see the water tension beginning to exert pressure?

Can you imagine what happens to a small insect that gets caught below the surface? If the surface is strong enough to hold the insect on top of the water, it is most likely strong enough to keep the insect from breaking through the top surface of the water, too.

 Another way to see water's surface tension requires two plates of glass or clear acrylic. Have an adult help you with this activity, which was first demonstrated in 1889 by C. V. Boys, a London scientist and lecturer. Fill a 9" × 13" pan with about a half-inch to an inch of water. Add a few drops of food coloring to make it easier to see. Take two pieces of glass or acrylic the same size. Put a wooden match or something of a similar size between the pieces of glass at one side. Put a large rubber band around both pieces. They will touch along one edge and be spread apart gradually by the match at the other edge. Put the bound plates into the colored water. Watch what happens.

The surface tension makes the water attract itself. Where the two surfaces of glass meet, the water "climbs" the edge. As the area between the surfaces increases, less water can climb.

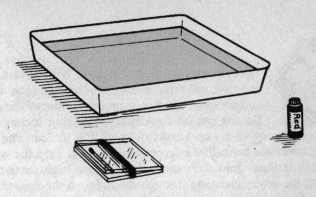

OBSERVING LIQUIDS IN SODA STRAWS

When you put a straw into a drink, does the liquid rise in the straw? Take a small sip. Observe again. Now that the straw has been wet inside, does the liquid rise a bit when you insert the straw? If you could gather several clear straws or tubes of different sizes, you would see that the liquid rises highest in the narrow tubes.

The liquid seems to climb up the inside of the straw because the skin of the water attaches itself to water inside the straw or between the plates of glass. The tension between the two liquids is greater than the air pressure and the gravity which normally cause the liquid to appear flat.

Knowing about surface tension can be helpful when you want to pour water from one container with a wide mouth to another container with a smaller mouth. How can you do it without spilling? If you use a straw or a stick and pour the water slowly down the side of the straw, the surface tension holds the water together in a small stream. It looks as if it hugs the straw.

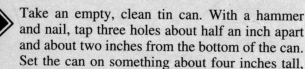 Take an empty, clean tin can. With a hammer and nail, tap three holes about half an inch apart and about two inches from the bottom of the can.

Set the can on something about four inches tall, like a coffee mug or an upside-down bowl. Put the can and the mug or bowl base beneath the faucet of your sink. Turn the water on so that it flows into the can. Just turn it on high enough to replenish the water escaping through the holes in the can. Is the water escaping in three streams? Pinch the three streams together. Remove your hands. What happens? Can you separate the streams again?

What held them together? Surface tension!

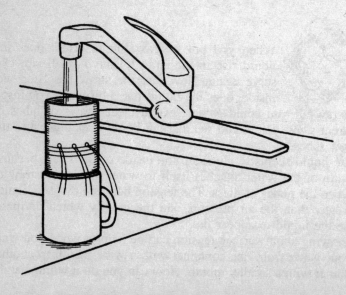

WHAT BREAKS SURFACE TENSION?

We know that gravity is a force that breaks surface tension. We've seen that we are strong enough to break the tension and put our hands into water.

Make a little boat out of a cork or nutshell. Set it in a container of water. Let it sit for a moment. Does it move? Wait until the boat is still. Now put a drop of dishwashing liquid* on one end of the boat. What happens to the boat?

*Note: In this book, dishwashing liquid refers to the liquid you use to wash dishes in the sink, not the detergent you use in your dishwasher.

 Here's another way to observe breaking surface tension. Lightly sprinkle flour or powder on top of a container of water. Then put a small amount of dishwashing liquid in one corner. What happens to the layer of flour?

 Get a clean bowl. Fill it half-full of clear water. Make a circle out of a piece of string about five inches long. Float the circle of string on the top of the water in the bowl. Now wipe your finger across a bar of soap or touch the top of the dish detergent bottle. Touch the inside of the circle with your soapy finger. What happens?

In these experiments, the dishwashing liquid weakens the surface tension. If we could see the molecules in water, we'd see that they attract each other because of electrical imbalances in their structure. A group of water molecules is very happy together. Remember how hard it was to break apart the drops of water on the waxed paper? Now imagine all these molecules reaching out in all directions, perfectly content. What about the molecules at the top of the pile? They don't have anything above them to pull in that direction. Therefore, the pull to the sides and down is much greater, creating surface tension.

Soap acts on the surface tension by supplying different molecules and charges. The surface tension isn't broken, just weakened. Water molecules are attracted to soap molecules.

BREAKING DOWN THE WALLS

If you turn on the faucet and let the water fill a pan, you will notice that bubbles form. If your pan does not have leftover soap from when it was washed, they will disappear very quickly.

Fill a bucket from the hose outside. Turn the water on slow. Then turn it up. Which makes more bubbles, fast or slow? Put your thumb across part of the hose end. You are creating pressure. Does the water bubble more or less?

In nature, we see these bubbles, too. Waves that break against the beach are white. And so is a river rushing over rocks or through a narrow canyon. We call that white water. Moving water mixes with air at the surface and bubbles are formed. They do not last long because the water pulls back, letting the air escape, but while the bubble is in the water, it reflects the sunlight differently than still or calm water.

The next time you are at a lake, watch different boats. Which makes more bubbles, which we call wake, the faster boats like a speedboat, or the slower boats that are trawling for fish?

The easiest way to do the activities in this chapter is to have a ring about eight inches in diameter. A plastic ring used in swimming games or a plastic macramé hoop works nicely.

Mix up some bubble solution by adding one-fourth cup of dishwashing liquid to two cups of water. As we'll discuss later, the more expensive dishwashing liquids make better bubbles.

If you don't have either a diving ring or a macramé or embroidery hoop, take two long drinking straws. Cut them so each one is about seven inches long. Thread some yarn through them, leaving about five inches of yarn extending from bottom and top of each straw. Tie the yarn together and trim the ends. Now you have a straw-yarn hoop.

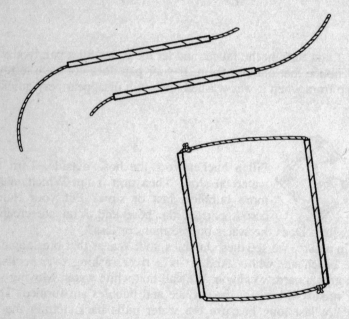

 Dip your ring into plain water. Do you get a film or skin of water across the ring? No, you don't. The water's surface tension is too great to allow the water to stretch across an eight-inch ring.

Dip your ring in your bubble solution. Does a film form across the ring? If it doesn't, try again. Make sure the entire ring is covered with solution. You should see a thin sheet of soapy water stretched from one side of the ring to the other.

Touch your film with your finger. What happens? Dip your finger into the bubble solution and touch a new film. Gently push your hand into the film until the film breaks. At what point does it break? Dip your entire hand into the bubble solution. Push your hand into a new film. Again, at what point does the film break?

What caused the film to break? Did a sharp fingernail break the film? Did a ring break it? Did dry skin break it? If you think it was a sharp fingernail that broke the film, ask to use a kitchen knife.

Make a new film on your ring. Touch the film with the knife. Now get a new film. Dip the knife into your bubble solution. Poke the knife through the film. Only insert the knife as far as you have bubble solution on the blade. Does the film break?

Turn on the kitchen faucet. Adjust it to a slow, steady stream. Dip your ring into bubble solution to make a film across it. Gently slide the ring across the stream. What happens? Adjust the stream so you can hold the ring in the stream without the film breaking.

 Take a piece of yarn and tie it across your ring. Dip the ring and get a new film. Break the film on one side. What happens? Break the other. What happens?

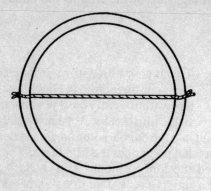

 Take another piece of yarn. Make a circle about an inch in diameter. Tie two pieces of yarn to the circle, on opposite sides, and then tie them to the ring. Dip this into your bubble solution. Break the film across the small circle in the center. What happens? Dip the entire ring again and break one of the outer films. What happens?

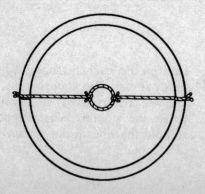

Take your ring made of straws and yarn. Dip it into the bubble solution and make a film. Hold this up in the air. Carefully twist one straw around ninety degrees while holding the other still. Let go of one straw and hold one. What happens to the shape of the film? Can you pull it back into shape without breaking the film?

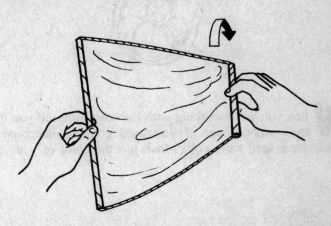

From these activities, you've learned that soap film has surface tension, too. When you broke half of the film, the other half contracted. When you twisted the yarn, the film contracted and didn't want to be restretched.

Surface tension of soapy water is essential to making bubbles.

BUBBLES AND RAINBOWS

When you were working with the soap film, did you notice the colors? Even if you used colored detergent, you should have seen more colors than just the color of your detergent.

Dip your ring into the bubble solution again. Hold it up and look at the colors. Do they stay the same? Do they change?

The colors in soap films, and in bubbles, come from reflected light. Even though bubbles and soap films are transparent, meaning you can see through them, they reflect some light. Most of the time that you're working with soap films and bubbles, you are surrounded by white light. Sunlight is white light. That means that all colors are possible.

When you break up white light, you get the full spectrum of colors, from red to violet, like a rainbow. A rainbow forms when white light (sunlight) is reflected by moisture in the air. Sometimes you will see a small rainbow when the sun shines on water from a sprinkler. You can see the spectrum of light if you use a prism.

Can you make a rainbow with a mirror in a pan of water? Have an adult help you. Try reflecting sunlight onto a white surface through the pan of water. Never, never, never look directly at the sun or reflected sun. It could hurt your eyes.

 Dip your ring into the solution. Watch the colors form and swirl. Hold the ring upright over your bubble solution container. Keep holding it and watching the colors as they change.

There are two surfaces of a soap film. They are so small we can't see them. It looks like one surface to us. The molecules of detergent line up on both sides of the molecules of water. The distance between the two is very small. Even so, it is enough to make the changing colors you see on bubbles and soap films.

The colors are caused by the interference of light waves. When light hits the outer surface, certain wavelengths of light are reflected. If the inner surface also reflects those same colors, we will see them. If the inner surface does not reflect that color, the two waves cancel out each other. For instance, if the outer surface reflects violet, but the distance between the two surfaces is such that the inner surface does not reflect violet, we won't see violet. This is negative interference. If both reflect violet, we'll see it. This is called positive interference.

This happens because light acts like a wave. Each color has a different wavelength.

As you hold your ring with a film across it, what happens? Eventually it will break. But before it breaks, the bubble solution changes in thickness. You can't really see it changing, but you can see that the solution builds up at the bottom and drips off. Where does this extra solution come from? Some comes from the edges of your ring. Some comes from the film itself. Gravity pulls at the solution. The film near the bottom gets thicker as the solution travels down. What does this do to the colors?

 Try holding your film in a beam of sunlight or bright light, such as a flashlight or light bulb. Can you see the colors moving? What color do you see just before the film breaks? When the solution drains away, the film becomes very thin. Eventually, it will become so thin that it will break. Do you know what color you see when the film is too thin to reflect light? If you said black, you're right. The absence of light is black.

How long does it take your film to thin enough for you to see a tiny black spot appear and the film to break? As we'll see later, this depends on many things. Often, we won't see the black color before the film breaks. Something else has broken it.

CHANGING THE COLOR OF BUBBLE SOLUTION

Try adding some red food coloring to your basic bubble solution. Do the colors seem different to you? Try mixing small amounts of bubble solution with several different food colorings. Do they make a difference? If you add several drops of color, your bubble may have a slight red tinge, but it will still reflect all the colors of the light around you.

What happens if you add a drop of vegetable oil or motor oil to the bubble solution? Does the oil affect the colors? Think about how a parking lot sometimes seems to have rainbows floating in puddles where a lot of oil has leaked out of people's cars.

 Ask to borrow some household cleaner with ammonia in it. Dip your ring into the basic bubble solution. Hold it over the container with solution in it. Have someone unscrew the top of the cleaner with ammonia. Be careful not to breathe the fumes. Hold the ammonia bottle near the film but DON'T POUR ammonia into the bubble solution. What happens to the colors? Move the ammonia away again. What happens to the colors now?

You've seen that the surfaces of a bubble film reflect light. You now know that white light contains all the colors and that colors have different wavelengths. When you make bubbles and play with bubble films, it's nice to know these facts. Even so, you can enjoy the variety and pattern of colors in soap films without understanding light waves and interference. The colors are just beautiful to see!

DOMES AND HOMES

Although these activities can be very messy, they're lots of fun. Check with an adult before you get started. You could do them outside on a picnic table. If you go outside, find a shady spot. You could do them inside in the kitchen. Either way, cover the table with an old plastic tablecloth. Make a dam around the edges under the cloth with rolled-up newspapers or a hose. Anything that will prevent the liquid from spilling over the edges is good. Don't forget to put down newspapers to catch the drips. There will be lots of them!

You'll also need at least a quart of bubble solution and something to blow the bubbles. For tabletop bubbles, some good blowers are plastic straws, plastic hose or pipe, cans with both ends removed, Styrofoam cups with the end removed, or a funnel. Try taping three soup cans together with duct or masking tape for long blowers.

What you're looking for in a blower is a long, hollow cylinder. It must be clean. Food particles or dirt will interfere with your bubbles. If you use cans, make sure they don't have sharp edges. As you learned from working with the soap films, the sharp edge probably won't hurt the bubble, but it might cut you.

You will need a smooth surface. You might want to use the table itself, if it is waterproof and smooth. Ask first! Don't do these tests on wood. They could stain the wood tabletop. It's probably best to cover the table with an old plastic tablecloth. Put rags or newspapers under the edges of the table to catch spills.

Gather everything you think you will need before you begin. In addition to straws and bubble blowers, you will want rags for mopping up unexpected spills, a plastic ruler, a piece of yarn about two feet long, a small plastic toy about an inch to an inch and a half tall, and a sheet cake pan or cookie sheet with sides.

 Pour bubble solution into the cake pan, so that it's about half an inch deep. Dip your bubble blower into the solution. Then, holding it just above the solution, gently blow some large domes, or hemispheres in the pan. A dome or a hemisphere is flat on the bottom and a half sphere above.

Take a straw and insert it into a bubble. Does it pop? Do you remember what you learned about popping soap film? What do you need to do to keep your bubble from popping when you insert the straw? Right! Make sure the straw is wet with bubble solution. If you keep a glass of bubble solution in the center of the table, you can keep your straws in it. You don't want to lay the straw down in bubble solution or some might get in your mouth.

Blow another dome. See if you can add enough air to make it cover the pan. Are you running out of air? Have a friend help you. If you both have cans, you can dip the cans into bubble solution and add a lot of air to a dome.

What is the secret to a long-lasting dome? Don't blow too fast. Blow a smooth, steady stream of air. Make sure everything the dome touches is coated with bubble solution. You may have to dip your finger in the bubble solution and run it around the sides of the pan to coat them. Otherwise, when you blow a large dome and it touches the sides of the pan, it will pop.

Blow a medium-sized dome. Dip your straw into the bubble solution and insert it into the dome. Dip it into more bubble solution at the bottom of the dome. Try blowing another bubble inside the dome. What would happen if you tried to blow another bubble at the outside edge of the dome? Can you blow a second bubble on top of the first dome?

Take your small plastic toy. Dip it into bubble solution. Using your bubble blower, make a dome over the toy. It is harder than it sounds. Keep trying. Will using a funnel that fits over the toy as your bubble blower help? If you make a large dome, can you put the toy in afterward? Try rolling a marble or wheeled toy through your dome. What happens?

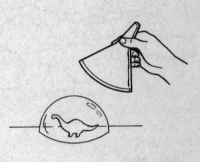

WHEN THE SIDE OF ONE DOME TOUCHES ANOTHER

Take a long piece of string. Dip it into bubble solution. Hold each end and slice through a dome. What happens?

Can you make a bubble "city" on the tabletop? Blow large and small domes. Make a "house" dome and see if you can attach a "garage" dome to it. Can you make an "apartment house" of connected domes?

Can you blow a series of connected domes to make a caterpillar? Can you blow small bubbles for its eyes?

Blow several domes together. Look at where they touch. What shapes do you see?

Can you break the wall between two touching domes? What happens to the shapes then? Carefully tear the wall between two touching domes with a straw. What happens?

Bubbles can teach us a lot about mathematics. Watch closely when you blow a dome. What happens when you add more air? Does it just get taller? Is there a relationship between the height and the length of a dome?

Blow a small dome. Using your straw, measure the length and height of the dome. Dip your straw into bubble solution and insert it into the tallest point of the bubble. Hold that point with your finger and lay it against a ruler. Write down the height. Lay the ruler next to the dome. Measure the length at the longest point. Write it down.

Blow and measure several more domes. Try to make them different sizes. Do you see a relationship between the height and length of a dome? Is the length about twice as long as the height?

Who would need to know this in the real world? Think about domes you've seen that aren't made of bubble solution. Have you ever seen a sports dome? A picture of the Capitol Building in Washington, D.C.? An igloo? A geodesic dome? People who build and people who design buildings know all about domes and hemispheres.

To Bubble or
Not to Bubble . . .

A soap film is a bubble without air. When you added air to the bubble solution in a pan, you made bubble domes. Let's make some other bubbles.

 Make a bubble maker from a pipe cleaner. Leave a couple of inches for a handle and make a circle with the rest. Close the circle and twist the wire around the stem to hold it in place. You can also use a wand from a commercial bubble solution. Dip your bubble

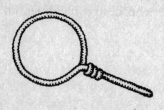

maker into the bubble solution and blow. Wave the wand through the air. Do bubbles form?

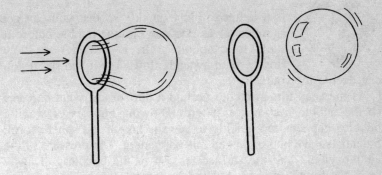

Do your bubbles travel upward in the air? Bubbles have a lot in common with hot air balloons. Your breath is probably hotter than the air surrounding you. When a balloon envelope is filled with air which is then heated, the balloon rises. Your warm breath should make the bubbles you blow rise.

When a balloon pilot wants to land, the pilot lets the temperature of the air in the balloon, or envelope, cool. The actual landing is trickier than that, but when the air cools, the balloon loses altitude. The air cools because the envelope is in contact with the cooler air surrounding it. In order to keep the balloon afloat, the pilot reheats the air with a propane burner.

Your bubble cools as well, and the air inside no longer can carry it aloft. Gravity overcomes the bubble, and it falls. If the bubble pops in midair, the bubble solution will quickly drop to the ground. If a balloon envelope tears, the balloon will fall. Good pilots know just how much air to let out for a smooth landing.

Did any of your bubbles seem to fly away? Air moves around in currents. We can't see them, but we can feel them. A balloon can hitch a ride on an air current and float far away.

When we blow a bubble, air becomes trapped inside the liquid until the liquid evaporates and becomes too thin to hold its shape. The surface of the bubble tears and the bubble pops.

 Is it better to blow quickly so that you can get as much air as possible into the bubble before it starts to evaporate? Or, is it better to blow slowly, with a steady flow? Try it and see if you can decide for yourself.

As the soap film stretches around a puff of air, what happens so the bubble becomes a sphere? Blowing bubbles or waving bubbles happens too quickly to see the last of the sphere form. Scientists have photographed this happening. The pressure of the air movement and the surface tension of the bubble pull away from the bubble maker. At the last second, the bubble forms a neck much like the end of a rubber balloon. This breaks away from the bubble maker and falls in on itself, closing the opening to form a sphere. If this doesn't happen, the soap film pops before it can become a bubble.

The pull of the surface tension is very great once the bubble forms. If a bubble lasts long enough, it will form a perfect sphere before it pops. Even extremely long bubble tubes will try to pull themselves together to form a sphere while dancing on air currents.

 Can you make bubbles with an electric fan? Try it outside. Hold a bubble maker in front of an electric fan set on low. Be careful not to touch the electrical connections with wet hands or get the fan near the liquid. Dip the maker into bubble solution. What happens? Keep trying, adjusting the distance between the fan and the bubble maker.

 Do this only with an adult. Try a hair dryer as your source of air. One of you will hold the hair dryer and the other will dip and hold the bubble maker. Be careful not to touch the electrical connections with wet hands or get the hair dryer near the liquid. What happens? Were you able to form bubbles with a hair dryer? Again, try adjusting the distance between the hair dryer and the bubble maker.

Which worked better, the fan or the hair dryer? Why do you think this happened? Bubbles hate heat because heat speeds up evaporation.

Bubbles also don't like carbon dioxide. When we inhale air, our lungs extract oxygen and give back carbon dioxide. The air we breathe out has more carbon dioxide in it than the air we took into our lungs. What does this mean for bubble making?

Try blowing bubbles with a pump or by waving your bubble maker in the air. Use the same bubble makers to blow bubbles with your breath. Do they last the same amount of time? Does your breath have enough carbon dioxide to make a difference?

Do this only with an adult. Is there air pressure in a bubble? You will need a burning candle and a funnel. Dip the large end of the funnel in your bubble solution. Blow through the small end, but don't blow hard enough for the bubble to detach. Hold your finger over the small end. Point the small end toward your burning candle. Remove your finger. What happens? How large a bubble does it take to blow out the candle? Keep trying until you find the perfect size. Why does the bubble blow out the candle? Remember that the surface tension is always pulling on the surface of the bubble, trying to make it as small as possible. When you uncover the hole in the funnel, the pressure forces the air inside the bubble out.

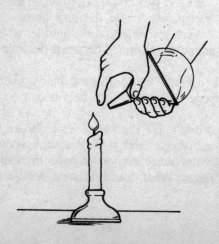

Use your funnel again. Wet the entire inside of the funnel. Blow a bubble on the large end. Cover the small end with your finger, then let go. Watch the soap film on the end of the funnel. What happens?

Which has greater pressure, a large bubble or a small one?

Take a margarine tub lid. Make sure it is clean. Punch a hole in the middle of it. Cover both sides of the lid with bubble solution. Take a straw and blow a bubble on each side of the lid, covering the hole. Make one bubble larger than the other. With your straw, break the soap film between the two bubbles. What happens? Which bubble pushed its air into the other one?

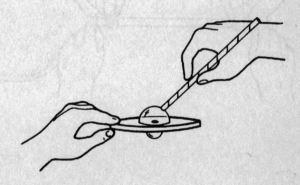

Another way to see this is with balloons. Balloons are like bubbles with plastic skins. Have a friend help you with this activity.

You will need a small piece of tubing, pipe, or hose, about six inches long. You will also need two balloons, the same size. Their ends should fit snugly over the hose. Blow each balloon, but don't tie them. Make one balloon about twice the size of the other. Hold the end of the balloon tightly closed, about an inch and a half from the end. Carefully slip the end onto the piece of hose. Do the same with the other balloon at the opposite end of the hose. You should still be clamping the balloon with your fingers. Have your friend hold the hose at the ends of the balloon to keep them on. Let go of each balloon at the same time. What happens? Which balloon exerted more pressure? Try this several times. What did you find out about pressure? Did the answer surprise you?

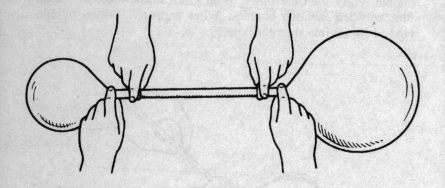

Blow two domes side by side. Make one larger than the other. Look at the shape of the film where they touch. Does it bulge slightly into one dome or the other? What shape is it? Can you tear the film to make just one bubble?

The rings we used to explore soap films make excellent bubble makers. You can blow into them or wave them in the air. Be sure your hand is coated with bubble liquid, too, so your dry skin won't break the bubble or the soap film.

Did you make a ring out of yarn and straws? Try making a larger one. Thread a yard to a yard and a half of yarn through two straws. Tie the ends together.

If you use a funnel, can you blow bubbles from both ends? How will they be different?

Can you design a bubble maker from a clothes hanger? You can bend a wire hanger into any shape you want. How can you get it to hold more bubble solution? You can't clip it like you did the plastic bottles. Think about the pipe cleaner. Its fuzzy edges hold the bubble solution. So does the yarn between the straws. Could you wind some yarn around the wire? Can you think of another way?

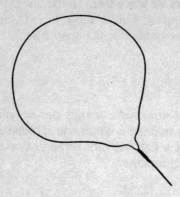

 Take a plastic or Styrofoam cup and poke a small hole in the bottom with a pencil or knife. Dip the rim of the cup into bubble solution. Blow through the hole in the bottom. Do you get large or small bubbles?

Can you think of other things to use for bubble makers? Think creatively as you look through your kitchen or garage. Just be sure you ask first!

 What could you use other than your lungs to supply air? You can wave a wand in the air. The movement causes air to rush through the ring, forming a bubble. How could you use a pump, like a bicycle pump or a turkey baster? Could you use an empty squirt gun?

Most toy stores have bubble toys. Science and museum shops have commercial bubble makers. A man named David Stein in-

You can use a bubble to power a boat. Fill your sink about half-full. Take a clean half-pound margarine tub. Make a hole in one side close to the top. This hole should be just the size of a straw. Insert the straw. Dip the open edge of the tub into bubble solution so a film forms across the top. Blow with your straw to make a dome above the tub. Hold the end of your straw closed with your finger. Set the tub on top of the water. Gently remove the straw. What happens? This activity is much like the one where you blew out a candle with a bubble on a funnel. The surface tension is exerting pressure all the time. When you remove the straw, the air is allowed to escape. It does so with enough force to power the boat.

Blowing bubbles is lots of fun. Are you ready to try some different bubble makers?

BUBBLE CONSTRUCTORS

Sometimes you'll want to try making huge bubbles. Sometimes you'll want thousands of tiny bubbles. There is probably something in your house that will make any size bubble you want to make.

You need to remember a few things. The bubble maker will have to have a closed shape, like a circle, through which you can blow. It can have more than one closed shape, like a figure-eight. It has to have a smooth opening to allow a soap film to coat it. Your bubble maker has to be something which can safely be coated with the bubble solution.

Make sure that when you borrow something, you have permission. And, always wash it when you're through.

Paper tubes from paper towel rolls make great bubble makers, but only for a short time. Why? Can you think of something that would last longer than paper?

Plastic pipe or hose will make a good bubble blower. So will clean, empty cans. The extra distance from your mouth when you tape' two or three together makes the air flow more smoothly. How would that be helpful in making very large bubbles? Did you try blowing fast versus blowing slowly? Were your slow-blown bubbles larger?

Plastic macramé and embroidery hoops are wonderful bubble makers. So are rings for which you dive into the swimming pool. Could you use a hula-hoop? How would you get the soap film on it? What kind of container would you need to hold your bubble solution for a hoop that big?

Try cutting the bottoms out of clean, empty plastic containers such as detergent bottles, bleach bottles, or shampoo bottles. Does enough bubble solution stick to the bottle to make large bubbles? If not, how could you help it to hold bubble solution? Remember that even with the addition of soap, the bubble solution has surface tension. It will try to cling to itself. When we worked with water and two sheets of glass, we saw that the water rose where the glass came together. Try clipping the plastic, about half an inch long, all around the bottom. The surface tension in the bubble solution should cause it to cling together at the places you clipped and more bubble solution will be available for you when you blow. Does it work? Will it help if you bend the edges?

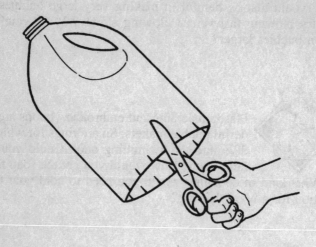

Can you think of any other way to increase the amount of bubble solution a plastic bubble maker will hold?

 Do you want thousands of tiny bubbles to float overhead? Try a piece of window screen or a plastic strawberry basket. A strainer will also give you lots of tiny bubbles. The plastic rings that hold a six-pack of sodas together will make several bubbles at once.

If you've had bubble liquid from the store, you may still have a wand. These work well. If you don't have one, you can make one out of a pipe cleaner or wire.

 Carefully curve one end around to make an open shape. Twist the end around the stem so it closes. Pipe cleaners work well because the fuzzy coating helps hold bubble solution. If you make a square bubble maker, will you get square bubbles? If you make it triangular, what shape bubble will you get?

People sometimes blow bubbles with bubble pipes. Don't borrow your father's pipe!

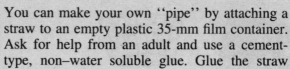

You can make your own "pipe" by attaching a straw to an empty plastic 35-mm film container. Ask for help from an adult and use a cement-type, non–water soluble glue. Glue the straw close to, but not touching, the bottom of the inside of the film container. When the glue is firmly set, pour about half an inch of bubble solution into the container and blow!

Straws are good bubble makers. They make small bubbles. What could you do to make larger bubbles with a straw? Think about what we did with the plastic bottles. Can you fringe the end of a straw and feather the fringed ends out? Why would that help?

The rings we used to explore soap films make excellent bubble makers. You can blow into them or wave them in the air. Be sure your hand is coated with bubble liquid, too, so your dry skin won't break the bubble or the soap film.

Did you make a ring out of yarn and straws? Try making a larger one. Thread a yard to a yard and a half of yarn through two straws. Tie the ends together.

If you use a funnel, can you blow bubbles from both ends? How will they be different?

Can you design a bubble maker from a clothes hanger? You can bend a wire hanger into any shape you want. How can you get it to hold more bubble solution? You can't clip it like you did the plastic bottles. Think about the pipe cleaner. Its fuzzy edges hold the bubble solution. So does the yarn between the straws. Could you wind some yarn around the wire? Can you think of another way?

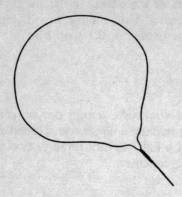

 Take a plastic or Styrofoam cup and poke a small hole in the bottom with a pencil or knife. Dip the rim of the cup into bubble solution. Blow through the hole in the bottom. Do you get large or small bubbles?

Can you think of other things to use for bubble makers? Think creatively as you look through your kitchen or garage. Just be sure you ask first!

 What could you use other than your lungs to supply air? You can wave a wand in the air. The movement causes air to rush through the ring, forming a bubble. How could you use a pump, like a bicycle pump or a turkey baster? Could you use an empty squirt gun?

Most toy stores have bubble toys. Science and museum shops have commercial bubble makers. A man named David Stein invented a bubble maker called the Bubble Thing. It makes HUGE bubbles. You can make your own bubble makers, but you also have one with you all the time.

 Dip your hand in bubble solution. Make an *O* shape with your thumb and index finger. Blow gently. You should get a beautiful round bubble that may want to sit in your bubble solution–covered hand like a crystal ball.

 Want a bigger bubble? Dip both hands into the bubble solution. Form a shape with both thumbs and index fingers. Blow through this shape. You may have to practice this. Remember, larger soap films break more easily than smaller ones, so you have to be patient. The results are worth the practice.

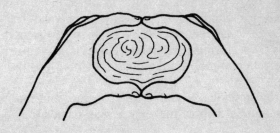

You need to have a container that will allow you to dip your bubble makers into it easily. For small- to medium-sized bubble makers, a plastic dishpan works well. If you have a small bubble maker and just a little solution, you can use a plastic cup. If you use a large bubble maker, you'll need a large container. Five-gallon plastic buckets work well with the Bubble Thing. A small wading pool works well with a large group of people or large hoops.

Remember to keep wiping away the froth that forms in the container. Bubbles don't like froth.

Cloudy with a Chance of Bubbles

When you've worked with bubbles a lot, you will notice that they are easier to do on certain days. Huge bubbles are not always possible.

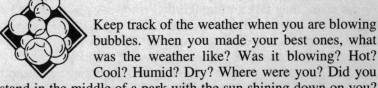

Keep track of the weather when you are blowing bubbles. When you made your best ones, what was the weather like? Was it blowing? Hot? Cool? Humid? Dry? Where were you? Did you stand in the middle of a park with the sun shining down on you? Were you under a tree in the shade? Had the sprinklers just been on?

When you pay attention to the weather, you will find there are conditions that make it easier to blow large bubbles and have them last longer. If you want to impress your friends with your gigantic bubbles, you'll want to plan your showtime around the weather.

You probably found windy days are not good for bubble making. The turbulent air interferes with launching bubbles and pushes them into dry objects which pop them. Dust is another thing that pops bubbles. A windy day may also be a dusty day if the weather has been dry.

A bubble pops when the water in it has evaporated. Days when the humidity is high are good for bubble making. The water in a bubble doesn't evaporate as quickly in humid air. At certain times of day the humidity is higher than at others. It is usually higher in the morning and evening than around midday.

Watch your local newspaper or weather forecast to see what time of day the humidity is highest in your area.

Cool days are better than hot ones. Heat encourages evaporation, which means death for your bubbles.

Some scientists believe bubbles are affected by the changes in weather and the ions that charge the air before a storm. After a rain is a great time to make bubbles. Before a storm is not as good a time.

In transpiration, plants give off into the air up to ninety percent of the water they take in through their roots. A place near a lot of trees and plants will be more humid than a parking lot. You will want to balance this with the fact that when your bubble touches the dry trunk or leaves of a tree, it will pop.

If you are standing near a building, you may notice air currents on one side of the building. If there is a breeze, try blowing bubbles on the side of a building away from the breeze. Updrafts will carry your bubbles up into the air, away from the ground and other dry things which will pop them.

If you live where it gets very cold in the winter, you can make a crystal bubble. Blow a small bubble and catch it on your bubble maker. Watch it carefully. Tiny crystals will begin to form. Eventually, the bubble will freeze completely, and you will have an ice crystal ball. It will be very fragile because the liquid surrounding a bubble is not very thick. You may have to practice a number of times before you get a complete ice ball.

How cold does the air have to be? Of course it must be below the freezing point of water, 32°F. But how much lower does the temperature have to be to freeze the bubble before it pops? Does the humidity make a difference? Try this activity and keep track of your successes and failures. Remember, you learn from failed experiments just as you do from successful ones. Keep trying!

What would be a perfect bubble day? A foggy day would be terrific because of cool temperature and high humidity, but would you be able to enjoy your long-lasting bubbles? Would they disappear into the fog? The best time to blow bubbles would be in the early morning or early evening of a cool day. The wind would be calm, and it would have rained recently.

We don't always have the perfect conditions. If you live in a dry climate, you will not be able to make bubbles as large as someone living in a humid climate. But it rains everywhere sometime. Take advantage of those times. When it hasn't rained, make your own humidity by seeking out a tree or by watering the grass before you begin. As we'll see later, you can also adjust your bubble solution.

Experiment and see what works best for you.

BUBBLE DIMENSIONS

When we worked with soap films, we used a ring. It gave us a single straight film. Can we make a frame that is three-dimensional? What would the film look like on a cube?

There are several ways to make frames. Choose one that calls for what you have available. Probably the easiest to make is by using plastic straws and paper clips.

Cut twelve identical straws to be the same length, about five inches. Attach two paper clips together. Slip one onto the end of a straw. Do this for all your straws. Attach two straws by slipping the free paper clip to the end of a different straw. Make two squares. Then attach four straws between the corners of the squares to make a cube.

You will need to have a container of bubble solution large enough to dip your entire cube in at once. A plastic shoe box, a dishpan, or an empty fish tank will work well. Try not to allow foam to form. If you get some foam, skim it off the top with your hand and wash it down the sink.

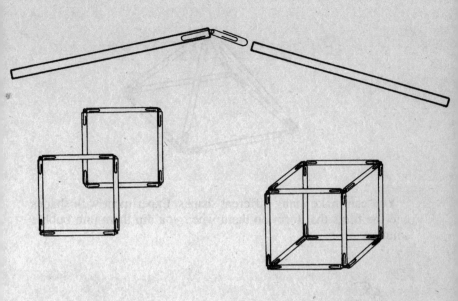

Dip your cube into bubble solution and raise it. What shape is the film? Does each side form a film? What happens in the center?

Remember that the surface tension will try to pull the film to the smallest shape possible. In the air, this is a sphere. Your frame changes the possibilities and the shapes.

You can make other shapes as well.

Make a pyramid with a square base and four straws meeting at the top. You will need to hook four paper clips together.

Make a triangle. Add three more straws, one at each angle, and fasten them together at the top. Each side should be an equilateral triangle.

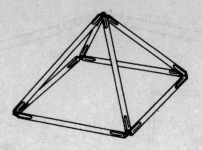

You can make many different shapes. Experiment with shapes and the films that form on them when you dip them into bubble solution.

Another way to make the shapes is out of wire. You will probably need some help with this. Be careful not to cut yourself. Cut twelve pieces of wire into equal lengths, six to seven inches long. At each end of a wire, make a loop, but don't close it all the way. You will want to hook the wires together with the loops.

You can also use pipe cleaners. Use them alone and twist the ends together at the angles you want. Or thread them through slightly shorter straws. They will have more stability with straws.

With wire or two long pipe cleaners attached together, you can make a spiral. Take a long piece of wire. Leave part of it straight for a stem to hold. Coil the rest of the wire around the stem in circles of increasing size. What kind of a film will form on this? Can you blow a bubble from a spiral? Use pipe cleaners to make other shapes. Make some with corners and some with curves.

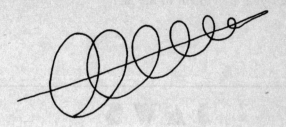

 Take a clean straw. Dip it into bubble solution. Insert it into the center of the film in a cube. Blow gently. What shape do you get? You may have to practice this several times.

 Use your straw to blow bubbles in other shaped frames. Where the films meet, try to form a bubble. What shapes will you get? Guess before you try it. Were you right?

 Can you make a rectangular bubble? Make a frame like the cube, but with four sets of four straws. Make at least one set longer than the other sets. Dip the frame into solution. Try forming a bubble in the center.

Try waving your frame in the air to see if you can make bubbles with it. Does it make many bubbles or just one? What shape bubbles do you get? Is it possible to get any shape other than round when you blow them in the air?

JAWS

One kind of bubble you probably blow a lot is a bubble gum bubble. Blowing soap bubbles and blowing gum bubbles are a lot alike. You have to stretch a film over an opening and force air through. With bubble gum, you use your mouth, tongue, and teeth to shape the film and blow the bubble through your lips. If the film is too thin, it will break and the bubble will pop.

What shape does a gum bubble form? The skin of a gum bubble is much like a soap bubble skin. They both want to pull back into the smallest shape possible.

Where does bubble gum come from? Gum is made from chicle, a rubbery substance that comes from the sapodilla tree. The tree is tapped much like maple syrup trees or rubber trees. Chicle comes out a tube that is inserted into the trunk of the tree.

Chicle is sent to a gum factory where it is mixed in huge vats with sugar and corn syrup. Flavoring is added. If the gum is to be bubble gum, some rubber is mixed in to make it stronger. Then machines roll the gum into sheets, cut the sheets into pieces, wrap the pieces, and put the pieces together into packages.

Bubble gum is usually pink because the first person to make it only had pink coloring available at the time. That was in 1928, and bubble gum has been pink ever since.

How big a bubble can you blow? The *Guinness Book of World Records* states that the largest recorded bubble was twenty-two inches, blown by Susan Montgomery Williams of Fresno, California, in 1985. How many pieces of gum do you think she had to use for a bubble that big?

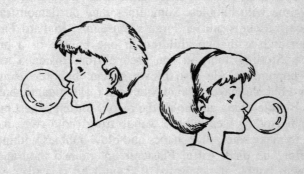

BUBBLE MAGIC

Bubbles are endlessly fascinating. Use your imagination and develop some tricks of your own. Or take the tricks here and expand on them. Can you make a bubble do something no one has ever done before? There are several people who spend a lot of time with bubbles. Sometimes they do demonstrations at science museums and children's museums. Richard Faverty makes an arch with two children in one bubble or a bubble doughnut. Sterling Johnson can escape from inside a bubble without popping it. He also blows special bubbles using his hands as a bubble maker. David Stein can make gigantic bubbles with his Bubble Thing and holds the record for the largest bubble ever. Tom Noddy uses bubble wands from dime store bubble solution to make caterpillar shapes and blow bubbles within bubbles. Before he died, Eiffel Plasterer entertained thousands of people with bubble tricks from 1932 until he was over eighty years old. He also enjoyed keeping bubbles in sealed jars to see how long they could last. One bubble lasted almost a year.

The following are some things for you to try with bubbles. Don't be discouraged if you can't do them the first time. Sometimes these things take a lot of practice. Sometimes they take near-perfect weather conditions. Even if they don't all work, you'll have fun.

Blow a large bubble and catch it on your hand. Dip your other hand in bubble solution. Can you make your hand look like it's inside a crystal ball? Remember that it has to be coated with bubble solution when you try to slip it inside the bubble.

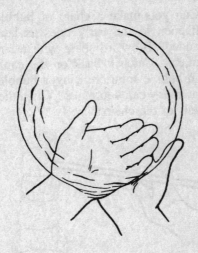

 Blow a medium-sized bubble. Look at it carefully. Do you see a reflection of yourself or the room in the bubble? Is it right side up or upside down? If things are reflected upside down, the image is being reflected off the concave, or inner, surface of the bubble. If you see something right side up, it's being reflected off the outer, convex, surface of the bubble. Take a teaspoon. Look at yourself in the bowl of the spoon. You're upside down. That's the concave side of a spoon. Turn the spoon over. Look at yourself in the back of the spoon. You should be right side up. That's the convex side of the spoon. Can you find both types of reflections in your bubbles? Make sure what you want to see is in plenty of light.

 Can you make a chain of bubbles? What will happen to the early ones as later bubbles are added? Blow a bubble with a soup can bubble maker. Catch it back on the can. Hold the can and dip a straw in bubble solution. Blow a bubble on top of the other bubble. How many can you make? When the oldest bubble pops, what happens to the chain?

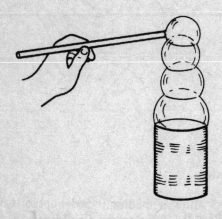

 You can also make a chain in a pan of bubble solution. Make your first bubble. Then make another bubble right next to it. Have them touch. Can you make a caterpillar? A ring of bubbles? Did your chain of bubbles in the pan last longer than those in the air? Why do you think this happened?

Take two bubble wands the same size. Blow and catch a bubble on one wand. Touch the top of the bubble with the other wand. The bubble should be between two wands. Can you make a cylinder out of the bubble by stretching it between the two wands? Will the ends ever flatten?

 Can you play catch with a bubble between two wands? Try it with a friend.

 This one is really messy. Do it outside, over a dishpan of bubble solution. Take a plastic Slinky. Dip it into bubble solution and start blowing a bubble through it. Keep blowing. What happens?

BUBBLES ON A PEDESTAL

For the next few bubble tricks, you will need to make bubbles and put them on a stand. Two possible stands would be a margarine tub of bubble solution or the top of a soup can bubble maker, dipped in solution. Some other things you will want to have on hand for these tricks include string or yarn, a piece of wool cloth, a wool glove, a plastic ruler, and lots of bubble solution and makers.

Blow a bubble, catch it with your bubble maker, and put it on the stand. Take a piece of wool and rub it against a plastic ruler many times. Hold the ruler close to the bubble. What happens? You've created static electricity with the wool and the ruler. What effect does static electricity have on a bubble? Try it several times. Then try this inside. Shuffle your feet along the carpet and reach out to the bubble. You've created static electricity in your body. Does it have an effect on the bubble?

Blow two bubbles about the same size. Catch them and put them on stands. Move the stands close together. Try to make the two bubbles touch. What happens?

Do this one with a friend. Take about a yard and a half of yarn or string. Tie the ends together. Dip it into bubble solution. Have your friend blow a bubble about six inches in diameter. Hold the string in both your hands. Stretch it out so it forms a track about three inches wide. Catch the bubble on the string. Can you make the bubble roll from one end to the other and back again? Trade places with your friend and try again. What happens if you have two bubbles on the string?

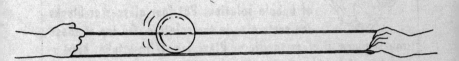

 Can you blow a bubble within a bubble? This trick takes a lot of practice and patience, but it's lots of fun to see when you're successful. There are two ways to do it, the harder way and the easier way.

The harder way is to blow a good-sized bubble using your hands as the bubble maker. Catch the bubble and hold it in front of you. Blow small, sharp puffs of air against the side of the bubble. Practice a lot. You should be able to blow fast enough to form a bubble using the original bubble skin for a film.

The easier way is to catch a good-sized bubble and blow another bubble inside it with a straw dipped in bubble solution. You have to be careful that the straw doesn't pop the original bubble.

BUBBLES AND BEEHIVES

Scientists have studied bubbles in clusters. They have found that a group of bubbles, all about the same size, creates a pattern much like a beehive.

Take a cake pan and fill it three-fourths full of bubble solution. Put four glasses or blocks the same size in the solution. On top of these rest a piece of clear acrylic or Plexiglas. With a tube, blow a number of bubbles under the glass. Try to make them about the same size. They should be too large to fit completely between the bubble solution and the glass. Their tops will be flat against the glass.

Cover your hand with a piece of wool or a wool glove (not one with man-made fibers). Blow a bubble and catch it with the wool. Does it stick? Does it pop? Try bouncing the bubble on your wool-covered hand.

Blow a large bubble and catch it in your bubble solution–covered hands. Can you stretch the bubble? How far can you stretch it? What happens when you stretch it too far?

Invent some tricks of your own. Can you pass a needle through a bubble? Of course! If you know the secret.

Watch how the bubbles shift and change in shape. How many sides touch? What size angles form?

Scientists have found that bubbles shift until three come together. Their angles are 120 degrees. Looking through the glass, the bubbles resemble a honeycomb. Bees form their wax with three sides touching and angles of 120 degrees.

Interestingly, this is a common pattern in nature. A turtle's shell, the brown pattern on a giraffe's coat, and dried mud all form this pattern.

VINCENT VAN BUBBLE

O ne of the joys of bubbles is that they are temporary, fleeting beauties sailing through our world for just a few seconds. Bubbles that last forever would cease to hold their fascination for us.

Even so, we look for ways to make longer-lasting bubbles. We'll learn about experimenting with bubble solution recipes in the last chapter.

Sometimes we also want to save something from the experience. Some people have photographs of their bubbles. Some write poems about bubbles. Some include bubbles in their artwork. Some even use the bubbles to create art.

If you want to photograph bubbles, you'll need to be aware of certain things. Professional photographers who capture bubbles on film for magazines use a lot of film. They are willing to shoot many frames to catch just the right shot. They also have to have willing subjects who will continue to make bubbles and do tricks with them. They know about lighting and can capture the iridescence of bubbles in their photos.

You can capture some of these things with almost any camera. Bubbles will show up best against a dark background. In contrast to the dark background, you will need light coming from the front and sides. Watch the bubbles for a while. Think about what you want to capture on film. Do you want many bubbles floating? One big one? A photo of the person blowing?

Ask yourself some questions about timing. About how many seconds are they lasting? How many seconds does it take for them to reach their best shape, the shape you want to capture? Once you've paid attention to these things, you will be able to judge when to take your photographs.

There are more ways to capture the fun of bubbles. If you add dye to the bubble solution, you can capture the dye when the bubbles pop. There are several ways to do this. These activities should be done outside, where the dye can be washed away easily.

 Add a little liquid tempera paint to your bubble solution. Add about a teaspoon for a quart of bubbles. Stir well. Blow some bubbles over a large sheet of plain white paper. When they pop, they will leave a "footprint" of the bubble. You may have to experiment with the amount of paint to add. Tempera can be thick and will interfere with the bubbles' flight.

Mix three different containers with bubble solution and a different color of tempera paint. Have two friends blow bubbles with you. Each of you can blow a different color. Catch the pops on paper. Which colors made the biggest bubbles? Which show up best on the paper?

 Try adding white tempera to your solution and catching the pops on black paper. Does it look like a wintry scene? Can you paint in other things to enhance your picture? Maybe it reminds you of space. Perhaps something entirely different comes to mind. There are no right answers, just lots of creative possibilities.

If you don't have liquid tempera paint, try a drop or two of food coloring. Again, experiment until you are happy with the colors you get.

 Take a jar of dyed bubble solution (either paint or food coloring). Cover a washable surface with layers of newspapers or an old plastic tablecloth. Have some plain white paper at hand.

Set the jar on top of a piece of paper. Blow bubbles with a straw in the jar, letting them foam over the top and sides of the jar. Wait until the bubbles have popped. Remove the jar.

 Another way to capture these bubbles is to blow bubbles in a jar of dyed bubble solution. Let the bubbles form and push each other to the top of the jar. When you have several at the top, care-fully slide the straw from the jar. Place a plain sheet of paper over the bubbles. Remove the sheet.

Do bubbles pop in circles? Do all pops look the same? Could you tell that the pictures were made by bubbles popping if you didn't know?

Try showing your pictures to someone who didn't watch you make them. See if she can guess how you made the pictures.

Still, you'd like to spend more time with your bubbles. Do you remember having your hair washed when you were little? Your mother or father probably made your hair stand up in funny ways while the shampoo was in it. They sculpted horns and spikes with the foam and your hair. There's a way to sculpt with bubbles, and you don't have to put it in your hair! Just use the Sculpting Bubbles recipe in the back of the book.

GIGANTIC BUBBLES

According to the *Guinness Book of World Records,* the largest bubble was a fifty-foot-long tube. It was made by David Stein. He's the man who invented the Bubble Thing, a wand for making very large bubbles.

Several things make his wand special. First of all, it is made from a fringy material that holds lots of bubble solution. This is important because while the bubble is forming, it needs a continuous supply of bubble solution. Think about it. A little solution can expand around enough air to make a little bubble. More solution means larger bubbles are possible.

Another thing that makes this wand successful is the size of it. The film across the opening is over two feet wide. Compare that with your wand from the commercial bubble liquid and you will see the advantage in the Bubble Thing.

The last thing that makes his Bubble Thing so successful is that it can be manipulated to open and close. This helps the large bubbles to form completely.

If you don't have a Bubble Thing, you can still make large bubbles. A large macramé hoop, about ten to twelve inches across, will make very large bubbles. You can also use a hula hoop or a covered wire hanger.

Dip the hoop into bubble solution. Be sure you have a film across the hoop. Your hands will get wet, too. Gently pull the hoop through the air.

You may need to experiment with breaking off the bubbles. Try twisting your wrist at the end, turning the hoop, and pulling the ends together a bit.

Remember the yarn and straws bubble maker? See how large a bubble you can make with one. Hold the soaked maker in your hands and pull it up through the air. This bubble maker has one of the advantages of the Bubble Thing. You can close the maker behind your bubble.

What if you and a friend made a bubble together? Each of you take one end of a straws and yarn bubble maker.

This time use about four yards of the fattest yarn you can find. Dip it in solution, making sure your hands get covered with solution as well.

Carefully step apart, stretching the yarn and creating a film. Together pull the bubble maker through the air. How large a bubble can you make? This might take a bit of practice, but the results should be worth it.

For really large bubbles you need a lot of bubble solution. You need a container that will hold your bubble maker to allow the film to form. If you're trying a hula-hoop, you'll need a small wading pool. If you use the straws and yarn, you can use a dishpan or plastic bucket.

Remember what you learned about weather. Even David Stein can't blow fifty-foot tubes in hot, dry air. And don't be disappointed if your early attempts are small. A two-foot sphere is still pretty impressive.

BUBBLES AT WORK

Bubbles are a part of our world. People blow bubbles with soap solutions and make them when they wash with soap. This is just a small part of bubbles in the world. Bubbles occur in nature. Bubbles are created in other ways than from soap. Bubbles can be useful, and bubbles can be destructive. As you'll see, some bubbles are useful by being destructive!

Bubbles give our soft drinks zip. A flat cola or root beer tastes horrible. Where are those bubbles in the bottle? When you look through the clear plastic sides of a pop bottle, you don't see bubbles. You see liquid. The bubbles seem to form when you pour the soda into a glass.

The bubbles in soft drinks are carbon dioxide, a gas. They are dissolved in the liquid while there is pressure on the liquid. That happens as the soda is bottled or canned. When you finally open the bottle or can, the pressure is released, and the gas bubbles up. After you drink the bubbly liquid, the bubbles are in your stomach. These same carbon dioxide bubbles make you burp.

The same is true of champagne and beer. If you drank a soda pop underground, say in the lunchroom at the bottom of Carlsbad Caverns, it might taste flat. There is more pressure below ground than above. When you came up, the gas would come out from being dissolved, and you would suddenly be filled with gas in your stomach and need to burp. However, if you took your container of soda pop that has lost its fizz to the top of Mt. McKinley, it might be fizzy again because the pressure is lower at high altitudes.

Scuba divers have to worry about air pressure and the gas dissolved in their bloodstream. The pressure below the water builds. Divers could not breathe were it not for their pressurized air tanks. The feeder valves on divers' tanks make sure they get air at just the right pressure to offset the tremendous pressure of the water. But while the divers are down, the pressurized air builds up in their blood. This isn't a problem until they start to rise.

A diver gets a condition called the bends if he rises too quickly and doesn't continually exhale. If he comes up too fast, he either has to go back down or decompress himself in a pressurized chamber.

Imagining air dissolved in a liquid is pretty hard to do. But think about how fish breathe. They filter water through their gills, gleaning from the water the oxygen that is dissolved there. You have to give your fish a tank which continually bubbles and introduces more air into the water, or you have to change the water daily.

Bubbles massed together to make foam are used extensively. Fire fighters use a foam which smothers a fire and cools it at the same time. Bomb squads can surround a bomb with foam which will absorb some of the shock of the bomb going off. This prevents much of the destruction that would normally occur.

It was suggested that embassies be set up so that if the embassy was attacked, the inhabitants could trigger bubbles to fill all the space in the building. The people who thought of this believed that the people inside could be trained to find their way out, and the attackers would be slowed considerably by having to push through foam and the fact that they would not be able to see where they were going.

Foam insulates our homes and our food. Some pioneers in the field of energy and house building have tried making homes out of Styrofoam and cement. A plastics company in Norway built eighty percent of a house from plastic. There are many uses for foams that we haven't even thought of yet. Maybe you will be a scientist and discover a use for bubbles and foam that will help people.

Most of the things we buy were not produced in our own areas. Goods have to be shipped. Many things can be put in boxes or cartons, but some things are very fragile. Have you ever received a package wrapped with plastic covered with bubbles? The sheets have thousands of bubbles, each encased in plastic. Together, they shield the fragile item from bumping and dropping during shipping. Next time you get a package with bubble wrapping, ask if you can have it. Then try popping the bubbles. Pop one at a time. See how loud you can make the pop. What happens if you wring the sheet in two hands?

A foam has been made from kelp, a type of seaweed, that is almost lighter than air and can support thousands of times its own weight. This foam, called SEAgel, would make a good sound barrier in aircraft or high-speed railways. It may also be used as insulation for refrigerators or oil tankers.

Scientists use bubbles, too. In 1952, an inventor named Donald Glaser was drinking a beer. He noticed that all the air bubbles released in the beer seemed to stream from the same place. He invented the bubble chamber. It is a device used by nuclear physicists to study the behavior of subatomic particles. A liquid, usually hydrogen, is heated beyond the boiling point but kept from boiling by intense pressure. When the pressure is reduced, electrons and other particles are shot through the liquid. They make strings of tiny bubbles, which show the path of the electrons.

Bubbles are formed when it rains on the ocean. The noise created by these bubbles helps scientists keep track of the amount of rain falling on various parts of the ocean.

Also in the ocean, large waves crash. They envelop a large volume of air which breaks up into many bubbles. These bubbles oscillate and give off a loud noise for a short time. Scientists are studying all the noises made in the ocean to be able to track submarines better.

Different scientists are working with old bubbles. These bubbles contain the air that was trapped in them long ago. Some of these bubbles are trapped in amber and contain air from the time of the dinosaurs. The scientists have found that the air in amber created millions of years ago contained much more oxygen than our air today contains. They think dinosaurs might have died out because the air was no longer oxygen-rich enough for them.

Air is trapped in other bubbles, especially bubbles in Antarctica, where they stay frozen. It will be interesting to see what scientists studying them find out.

Doctors are using bubbles to help in their work. Two researchers found that their "bubbicles" or long-lasting, air-filled particles help doctors see tumors in the livers of their patients. The cells of the liver absorb the bubbicles, but the cancerous cells do not. The ultrasound transmitter sends out sound signals that reflect off the tissues in the body. Then the echoes are picked up by the receiver. These echoes are electronically sorted and an image emerges. The cells of the liver are brighter than the dark cancer cells.

Doctors use bubbles to destroy kidney stones, too. A spark is introduced into the kidney by a laser. This creates a gas bubble. The bubble collapses and creates a shock wave that breaks up the stone. This process is called cavitation. It is an example of a bubble that is destructive and helpful at the same time. It destroys the kidney stone and helps the doctor and patient.

Bubbles can get in your paint and make little pits where they pop on the wall. You have to paint over them to cover them up.

Bubbles also cause pitting on machinery that comes in contact with moving fluids. Turbine blades, pumps, and propellers are some of the machinery that is damaged by bubbles. Some scientists say the damage is caused when the bubble is formed, but others say it occurs when the bubbles pop. More studies have to be done.

Bubbles are all around us. They are in the kitchen sink, helping take the grease and dirt off our dishes. They are in the bathtub, helping scrub away dirt and leaving our skin clean. They work hard. They are also fun to play with.

BUBBLE EXTRAVAGANZAS

One way to share the excitement and joy of making bubbles is to have a special day of bubble making. You might want to have a neighborhood bubble day in your yard. You might prefer to get your class at school involved. Maybe even your whole school would like to participate.

Before you invite everyone, you'll have to do some planning. To get you started, you'll need to think about the following questions.

- Where is a good place to hold the bubble day? This will partly depend on how many people you expect. It will also depend on what is available to you without charge. Remember some of the things you learned about bubbles and weather. Bubbles like the shade because direct sun quickly dries them out. Bubbles like updrafts. Windy locations are not good. Big bubbles need space to form and float.

- Who will supply the bubble solution and bubble makers? You can easily make a gallon of bubble solution for yourself and a few friends, but what if you expect more people? You might have everyone bring his or her own, if everyone is familiar with the recipes. You might have members of your class or neighborhood sign up to bring certain ingredients or bubble makers. Share the responsibility of planning a big event. Bubble solutions seem to work best if they have at least twenty-four hours to sit before you use them, so you need to plan your day in advance. It will be important that there be plenty of bubble solution, too. Provide enough bubble makers so everyone can have one. Do you want just one kind? Do you want to have a variety?

- Whom will you invite? Remember that someone will have to watch little children to be sure they don't taste the bubble solution. Someone will have to make sure they don't suck in on a straw instead of blowing. Do you want to invite just kids your age? Do you want to invite little kids, but say they have to bring along an adult?

- What will you do about uninvited people who show up? If you have a bubble bash at a local park or school yard, you will have bystanders who will want to join you. What can you do to be prepared for them?

- Will your bubble program be a demonstration or show, or will you have contests and prizes? If you want to dazzle an audience with your bubble wizardry, you will need to practice. A lot. After your show everyone will want to try what you have done. Will you have a place and supplies for them to try it?

- If you have a contest, what will be the criteria for winning?

- What categories will you judge? Who will be the judges?

- Will you have categories for different age levels?

- How many categories can you include in the time you've allotted?

- What will you do if it rains? What if it is very windy?

- If you have contests, what will you give for prizes? Can you make ribbons or certificates? Be sure you let people know in your invitations or advertising just what the prizes will be.

There are a lot of things to consider. Don't let them keep you from putting on a bubble day. It is probably a good idea to start small. Invite three or four friends to make bubbles in your backyard. You can also get some adults involved in the planning. Your teacher might be willing to have a bubble day be a class project. The librarian at your branch library might be willing to help. Be sure to check things out with your parents before you make any plans.

The following is a list of possible categories of bubbles to judge. Decide which appeal to you and your friends. Add ideas of your own. Don't try to do them all unless you have a lot of help.

- Who can make the largest bubble?
- Who can make the smallest bubble?
- Whose bubble lasts the longest?
- Who can blow a bubble within a bubble?
- Whose bubble is closest in size to a baseball? A basketball?
- Whose bubble looks the most like a football?
- Whose bubble is the most beautiful?
- Which bubble solution is best for large bubbles?
- Which bubble solution lasts longest?
- Who can put a person in a bubble?
- Who can make the most bubbles in one blow?
- Whose bubble floats the highest?
- Who can be first to blow a bubble and catch it?
- Who can be first to put a dome over a toy?

You can also have sculpting prizes if you have a place for sculpting. Be sure you've tried the recipes and know what to do.

- Which sculpture has the best design?
- Which sculpture is the most unusual?

- Who made the best creature?
- Which sculpture reflects the most colors?

An extension of your bubble day might be a photo contest of pictures taken on your bubble day. Be sure to have a camera around just to record the fun.

Whatever you do, be sure you have planned ahead well. It should be fun for everyone, especially you.

BUBBLE CHEMISTRY

The following pages give you suggestions for different types of bubble solutions. All work well. None of them is perfect. You will need to practice with them and decide which you like the best. You can also individualize them by adding ingredients in varying amounts. You will experiment. When you find a recipe that works well for what you want to do, be sure you write it down.

One of the things you'll have to vary is the water to liquid dish detergent ratio. The more detergent you use, the soapier your bubbles will be. This should make them longer-lasting. However, if you live in a dry climate or if your normally damp climate is unseasonably dry when you want to make bubbles, you'll have to add more water. If you don't, the air will quickly dry out the water in your bubbles and they will pop.

It's a good idea to keep a bubble notebook. In this notebook, you can write down which recipe you used and how it worked that day. Make notations about the weather: hot, dry, windy, foggy, just rained, cool, sunny, cloudy, etc. Then note which formula you used for your bubble solution. What did you add?

Then you can make notes about the kind of bubbles you were able to get. Was it a good formula for medium-sized bubbles that floated off on air currents? Was it too dry for bigger bubbles? Were you able to make any giants?

Were you trying to make a bubble within a bubble? Was the formula you used good for that trick?

What is the best formula for using your hands as a bubble maker?

What formula is easiest to whip up when friends drop by? Are you able to use any of the formulas right after mixing?

Make notes about your bubble makers. Which work best for you? Which make the largest bubbles? Which make the best size for bubble tricks?

If you keep a bubble notebook, you may discover the best recipe for large bubbles in your area and climate. You may even challenge David Stein's record for producing the largest bubble!

Here are some suggestions for experimenting with the formulas.

- Use small quantities for experimenting. A half cup of bubble solution will allow you to make small- to medium-sized bubbles with a commercial or pipe cleaner bubble maker.

- Write down your own formulas. You might discover the perfect solution for your weather or climate. You will want to be able to do it again.

- Don't limit your experiments to what you find listed. Think about the suggestions. What is similar to an ingredient listed and might be a good thing to try?

- Be sure you ask before you borrow something from the kitchen or garage.

- Clean up after yourself.

- Be creative.

- And, most of all, have fun!

Please note: The dishwashing liquid referred to in the formulas means the liquid you use to wash dishes by hand, not detergent used in the dishwasher. Detergent made for dishwashers is formulated *not* to make bubbles and foam. We have found that

Dawn, Joy, and Ajax work well in bubble formulas. Which works best for you?

Try to avoid making foam when mixing the formulas.

Most bubble formulas will work best if you let them sit twenty-four hours, or at least overnight.

Basic Bubble Formula #1

¾ cup water
2 Tbsp. dishwashing liquid

➤ THIS can be mixed in a clean jar. Measure and add the water to the jar. Measure and add the dishwashing liquid. Stir the dishwashing liquid into the water carefully.

· · · · · · · · · · · · · · · · · · · ·

Basic Bubble Formula #2

½ cup dishwashing liquid
8 cups water

➤ USE a container that will hold more than two quarts, such as a dishpan or clean gallon-sized milk jug. Measure the water into the container. Add the dishwashing liquid. Carefully mix the two together.

· · · · · · · · · · · · · · · · · · · ·

Basic Bubble Formula #3

6 cups water
2 cups dishwashing liquid
1 cup glycerin

➤ USE a container that will hold at least two quarts. Measure and add the water. Measure and add the dishwashing liquid and glycerin. Carefully mix the ingredients together.

➤ YOU will find glycerin at a drug store.

· · · · · · · · · · · · · · · · · · · ·

Basic Bubble Formula #4

½ cup dishwashing liquid
¼ cup glycerin
4½ cups water

➤ USE a quart-sized container. Measure and add the water first. Then add the glycerin and the dishwashing liquid.

➤ CAREFULLY stir the ingredients together. Try not to make foam.

. . . .

Basic Bubble Formula #5

1 cup dishwashing liquid
8 cups water
4 Tbsp. glycerin

➤ USE a container that will hold more than two quarts. Measure and add the water. Measure and add the glycerin and dishwashing liquid.

. . . .

Basic Bubble Formula #6

4 cups dishwashing liquid
1 Tbsp. glycerin
1 drop motor oil
¼ cup dark corn syrup
3 drops food coloring

➤ THIS can be made in a quart jar. Measure and add the ingredients. Stir carefully. (Corn syrup can be found in the baking aisle of your supermarket.)

➤ NOTE: This makes a concentrate. Mix one cup of concentrate into twelve cups of water.

· · · · · · · · · · · · · · · · · · · ·

Basic Bubble Formula #7

½ cup glycerin
¼ cup dishwashing liquid
2 Tbsp. corn syrup

➤ MEASURE and add these ingredients together. Stir carefully.

· · · · · · · · · · · · · · · · · · · ·

Basic Bubble Formula #8

½ cup Ivory soap flakes (for washing clothes)
8 cups hot water

➤ USE a container that will hold more than a quart. Pour in the soap flakes. Add the hot water. Stir. Let stand for three days. Stir in one tablespoon of sugar or honey.

....

Shampoo Bubbles

2 Tbsp. shampoo
½ cup water

➤ ADD the shampoo to the water. Stir. You will get small to medium bubbles with shampoo. Notice the swirling of the bubble solution on the bubble surface. You will have to stir this often as you use it.

....

Bubble Bath Bubbles

1/4 cup liquid bubble bath (Mr. Bubble works well)
1/4 cup water
1 Tbsp. dishwashing liquid

➤ STIR the bubble bath into the water. Add up to one table-spoon of dishwashing liquid as necessary.

....

Sculpting Bubbles

1/2 cup hot water
1/2 envelope unflavored Knox gelatin powder
1/2 cup bubble solution

➤ SPRINKLE the gelatin powder over the hot water. Let it sit a few minutes. Stir. Put it into a very large mixing bowl. Add the bubble solution.

➤ ASK for help and use an electric mixer to whip lots of air into the solution. Mix for about two minutes. Then turn off the mixer and lift the beaters. Do peaks of whipped bubbles keep their shape when you lift out the beaters? If not, mix some more.

➤ WHEN you have a stiff solution, try sculpting the foam into shapes on a cookie sheet with edges or a cake pan.

....

Experimenting

Here are some possibilities for you to try.

- Try using distilled water instead of water from your faucet in any of the recipes. Does it make a difference? Do you know what your tap water has in it that distilled water does not contain? (Minerals and other impurities.)
- Try using different dishwashing detergents. Which works best for you?
- Try adding any of the following in small amounts, about one teaspoon to a cup and a half of bubble solution:

  - Jell-O powder
  - Certo
  - honey
  - light corn syrup

  - dark corn syrup
  - pancake syrup
  - Corn Huskers lotion
  - sugar